YOUR KNOWLEDGE HAS VALUE

- We will publish your bachelor's and master's thesis, essays and papers

- Your own eBook and book - sold worldwide in all relevant shops

- Earn money with each sale

Upload your text at www.GRIN.com and publish for free

Imprint:

Copyright © 2017 GRIN Verlag
Print and binding: Books on Demand GmbH, Norderstedt Germany
ISBN: 9783668811799

This book at GRIN:

https://www.grin.com/document/444170

Adeyemi Phillips

Resource requirements and multiple values of biogas technology for rural households in developing countries

GRIN Verlag

1.0   INTRODUCTION

In the last few decades, increasing emission of greenhouse gases worldwide has become a major concern. With a growing world population and with increasing energy consumption coupled with higher living standards, there is a huge challenge in limiting the emissions of pollution and greenhouse gases. Moreover, the available fossil fuels are limited; consequently, with increasing prices we need alternative energy sources to replace the fossil fuels. The European Commission has set the goal that by 2020, 20% of the energy consumed should come from renewable energy sources, as well as 10% of the energy consumed within the transport sector [EC, 2011]. The commercially renewable vehicle fuels available today are ethanol, biogas, biodiesel, and electricity produced from renewable energy sources.

1.1   BIOGAS

Biogas is a renewable, high-quality fuel, which can be produced from a lot of different organic raw materials and used for various energy services. Biogas technology has been developed and widely used over the world, because it has a lot of advantages, including reduce of the dependence on non-renewable resources, high energy-efficiency, environmental benefits, available and cheap resources to feedstock, relatively easy and cheap technology for production, extra values of digestate as a fertilizer, etc. But the current status of biogas production and utilization largely varies among the different continents.

Biogas is produced when microorganisms degrade organic materials in the absence of oxygen. This process is also named anaerobic digestion (AD). The feedstock can derive from the agricultural, industrial or municipal sources. To date, in order to obtain a higher biogas yield, a lot of agricultural biogas plants digest manure with some additional co-substrates for increasing the content of organic materials. Besides input materials, biogas yield and AD process are affected by several other factors. There are a lot of different types of biogas plants over the world, and they are accepted and widely used by different countries. For example, floating drum and fixed dome biogas plants are two major types of small to medium scale biogas digesters used in African countries.

The implementation of biogas technology provides benefits in terms of positive environmental impacts and additional values of digestate used as fertilizer if considering current energy consumption, waste handling and agricultural production practices. In addition, biogas itself can be used in several ways: either raw or upgraded, such as production of heat or steam (the lowest value chain utilization), electricity production with combined heat and power production (CHP), upgraded and utilization as vehicle fuel, upgrading and injection in the natural gas grid. There are big differences of biogas utilization among various countries, in particular between developing countries and developed countries. In spite of the multiple benefits of biogas systems, present biogas production only uses a small part of the potential.

1.2   AIM AND OBJECTIVES

Therefore, this seminar report is aimed to review the production and application of biogas, based on the following objectives:-

1) To review the study various aspects of biogas technology, including its production, feedstock, different types of digesters, etc

2) To review the benefits of biogas technology, including the energy value (biogas utilization), environmental benefits, and the values of digestate; its installation costs and economic performance.

CHAPTER TWO

2.0   LITERATURE REVIEW

The global energy demand is increasing rapidly, and about 88% of this demand relies upon fossil fuels to date (Weiland, 2010). The energy demand will continue to grow during this century. However, GHGs emissions have become one of the most severe environmental problems. Use of fossil fuels is one of the main reasons for these emissions. According to the report of Intergovernmental Panel on Climate Change (IPCC), GHG emissions must be reduced to less than half of global emission levels of 1990 in order to minimize climate change impacts and global warming. Besides, the energy supply is another important global challenge, because some continents such as Africa are already faced with an energy crisis but most of the known conventional oil and gas resources are concentrated in politically unstable regions.

Today, there is a lot of research focusing on renewable energy resources. The development of renewable energy technology can help to reduce the dependence on the non-renewable resources and the problems of environmental degradation related to fossil fuels (Parawira, 2009). Biogas which is a renewable energy resource from wastes, residues, and energy crops will play an important role in future. The production of biogas from anaerobic digesters has significant advantages compared with other forms of bio-energy production. Firstly, biogas production has been considered as one of the most energy-efficient and environmentally beneficial ways to produce renewable energy. Secondly, it can use locally available and cheap resources to produce biogas, and it drastically reduces GHGs emissions compared to fossil fuels. Thirdly, the digestate associated with the biogas production is considered as an improved fertilizer that could partly substitute for mineral fertilizers.

2.1   Biogas Technology Status in Africa

The African continent has already encountered an energy crisis, including both commercial (petroleum products, natural gas, coal, and electricity) and traditional energy sources (wood and other biomass) (Parawira, 2009). However, the energy consumption and demand of the African continent is estimated to grow continuously, at rates even faster than developed countries. The factors contribute to this increase include the growth in population, energy demands from various domestic sectors and the demand for improving quality of life. In order to meet the Millennium Development Goals (MDGs), especially MDG1—reducing by half the percentage of people living in poverty by 2015, it is required to improve the quality and magnitude of energy services in developing countries (Parawira, 2009). In eastern and southern Africa it is estimated that energy use significantly relies on traditional biomass energy technologies but hardly takes modern, sustainable energy technologies. Due to the current economic situation in most African countries and the shortage of commercial modern energy, it is almost unlikely that the fossil fuels substitute for biomass (Parawira, 2009). The fossil energy resources distribute on the African continent unevenly, which leads 70% of countries in Africa rely on imported energy resources (Parawira, 2009). Certainly, biomass is an inexpensive and abundant resource, but if used in an inappropriate and unplanned way it will limit regenerative utilization and cause significantly environmental consequences. So it may be helpful to change the energy situation in Africa in ways of upgrading the biomass to higher-quality energy carriers.

The problems of traditional biomass fuels and non-sustainable fossil fuels have caused widespread research on the production and application of new and renewable energy resources, such as biogas, bio-fuels, and biodiesel. It is necessary to develop the renewable energy technologies, in particular biogas technology, because it helps to reduce the dependence on non-sustainable resources and the environmental degradation problems caused by the fossil fuel. Compared with other renewable energy production systems such as biodiesel and bio-ethanol, biogas production systems are not complicated and can be built and operated at both small and large scales in urban and rural areas. Moreover, the biogas technology does not compete with food production but biodiesel and bio-ethanol technologies do (Parawira, 2009). According to global experience, biogas technology is a relatively simple technology in term of the requirements of construction and management. It has been considered as a appropriate, adaptable and locally acceptable technology in Africa (Parawira, 2009).

Various international organizations and foreign aid agencies have made a lot of efforts through their publications, meetings and visits to promote the biogas technology and stimulate the interest of biogas technology in Africa. To date, some digesters have been constructed in several sub-Saharan countries. Various wastes are used as feedstock for biogas production, such as wasters from slaughterhouses, agricultural wastes, industrial wastes, animal dung and human excreta. The exact number of plants installed in Africa is unknown but most plants were installed in Tanzania and Kenya. In other African countries only a few up to hundreds biogas plants have been installed (Van Nes and Nhete, 2007). However, most of biogas plants installed in the African continent are small-scale plants, and the development of large-scale AD technology in Africa is still embryonic. Unfortunately, it is estimated that 60% of plants installed in Africa failed to stay in operation, although other plants show the success in providing benefits to the users over a number of years and the evidence on the reliability of the technology if properly operated (Van Nes and Nhete, 2007). In most cases, in order to promote the biogas technology some demonstration projects were introduced usually free of cost by governmental structures. It is assumed that the demonstrated benefits of running the biogas plants would stimulate people to adopt this technology automatically. However, it seems that this approach has not caused widespread promotion and the market of biogas technology failed to develop. Moreover, most of the installed plants are abandoned eventually.

Generally speaking, the government expects to disseminate the biogas technology over Africa based on a market-oriented approach, but it has not achieved to date. An only exception may be Tanzania, where most of the plants have been installed on a semi-commercial basis, but a large-scale dissemination is still not achieved (Van Nes and Nhete, 2007).

There are a number of constraints that affect the implementation of the biogas technology on large scale in Africa, including (Parawira, 2009):

1. Inexperienced contractors and consultants leading to poor-quality biogas plants and poor choice of materials;

2. Lack of reliable information on the potential benefits of the biogas technology;

3. Lack of academic, legislation and commercial infrastructure in the region;

4. Lack of knowledge on the biogas system in practice; -

5. Poor ownership responsibility by users;

6. Lack of pilot studies and full-scale experience;

7. Lack of properly educated operators and technical knowledge on maintenance and repair;

8. Poorly informed authorities and policy makers;

9. Failure to support biogas technology through the energy policy by government;

10. Research at universities is sometimes considered to be too academic in practice.

2.2   Types of Biogas Plant in Africa

There are numerous types of biogas plants over the world, categorized according to the type of digested substrates, according to the technology applied or according to the plant scale, etc. Briefly, a biogas plant has to consist of two components: a digester (or fermentation tank) and a gas holder. Usually the digester is a cube shaped or cylindrical waterproof container including an inlet which introduces the fermentable mixture in the form of slurry into the digester. And the gas holder is an air tight steel container which cuts off air from the digester and collects the gas produced and it normally floats like a ball on the fermentable mixture.

There are different types of small to medium scale biogas digesters which have been developed in African countries, including the floating drum, fixed dome, and plastic bag design. The former two have been applied widely in Africa. The fixed dome digester and the floating drum digester are shown in Figure1. The major differences between the two digesters are the gas collection method, which the gas holder of the fixed dome type is equipped with a gas outlet and its digester has an overflow pipe to lead the sludge out into drainage, but the digestion processes of the both two digesters are the same (Amigun and Blottnitz, 2007).

Table 1 shows the comparison of constructed material, capital investment, output, life time and advantages/disadvantages between these two types of biogas plants.

Depending on the text, any type of biogas plant may be used. Nevertheless, most of the plants installed so far are the fixed dome type in Africa because of its advantages. There are no moving parts designed for the fixed dome type and also no rusting steel parts existing so a long life of the plant, 20 years or more, can be expected (Amigun and Blottnitz, 2010). The biogas plant is constructed underground which can protect it from physical damage and save space.

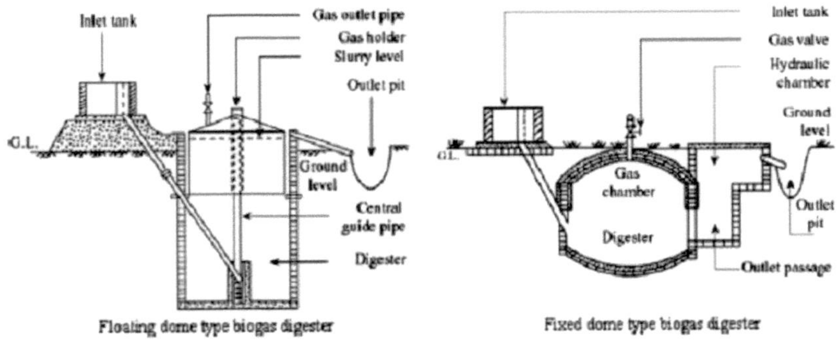

Figure 1:- Typical biogas plants of floating drum and fixed dome (Amigun and Blottnitz, 2007)

Table 1:- Comparison between fixed dome and floating drum biogas plant

Type of Biogas Plants	Constructed Material	Capital Investment	Output	Life Time	Advantages/Disadvantages
Fixed Dome	Locally available materials, which even could be bricks	low	low	long	A longer life(20 years or more); Easier to construct; Lower costs of installation, annual operation, maintenance
Floating Drum	Concrete and steel	high	low	short	Changeable space of gas storage; Less risk of uncontrolled gas outflow due to steel gas cover

Maintenance is required as occasional inspections, and if necessary, repairing the pipes and fittings. But the installation itself needs limited maintenance if operated properly.

A lot of studies have shown the technical and economic feasibility of fixed dome biogas plants. For instance, the fixed dome biogas plants are considered as technically suitable in Nigeria because they are easy to be constructed and the maintenance costs lowly (Amigun and Blottnitz, 2010). The economics of family size biogas plants of floating drum and fixed dome type in Punjab, India, with capacity between 1 and 6 m3, were compared by a research, and found that the fixed dome biogas design was the cheapest model as far as the cost of installation, annual operational cost, and payback period is concerned (Amigun and Blottnitz, 2010).

2.3 Biogas Production Process, Feedstock, Working Conditions

2.3.1 Biogas Production Process

Biogas is produced by biological processes which occur under anaerobic conditions. Biodegradable organic materials are mainly converted into methane (CH_4), carbon dioxide (CO_2) and small amounts of hydrogen sulphide (H_2S), moisture and siloxanes by anaerobic microorganisms. The process typically runs in a closed reactor at elevated temperatures or digester without heat system in the absence of oxygen. Nevertheless, it also could occur naturally in soils or old landfills at ambient temperatures (Omer and Fadalla, 2003). The degradation is a complex process, which requires some certain conditions and participation of different bacteria populations. The anaerobic fermentation processes are briefly shown in Figure 2.

The mixed bacterial populations degrade organic compounds and produce a valuable mixture of gases (biogas). The organic compounds undergo three main reactions which are hydrolysis, acetic acid formation and production of methane.

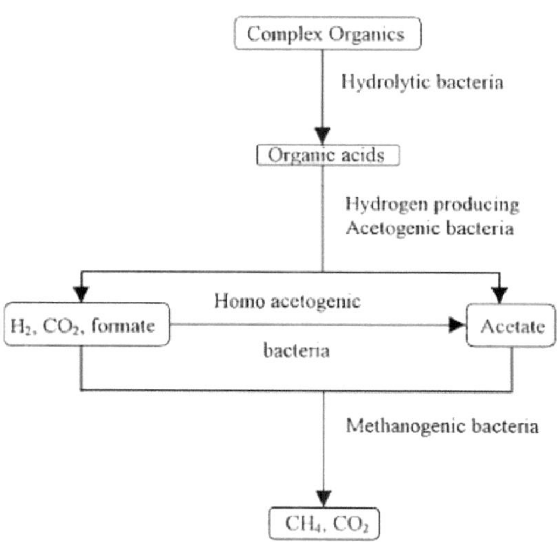

Figure 2:- Biogas production process (Omer and Fadalla,2003)

- **Hydrolysis**

Hydrolysis is a process that organic macromolecules such as carbohydrates, proteins and fats are de-polymerized by extra-cellular enzymes, then producing the acetic acid, long chain fatty acids and CO_2 (Lastella *et al.*, 2002).

- **Acetic acid formation**

Different bacteria degrade long chain fatty acids, then producing acetic acid, molecular hydrogen and CO_2 (Lastella *et al.*, 2002). Acetic acid can be produced from CO_2 and H_2, fatty acids, alcohols and carbohydrates (Lastella *et al.*, 2002). Enzymes for such reactions are named acetogens.

- **Production of methane**

Acetic acid is finally degraded, then producing methane by the so-called methanogenic bacteria or methanogens, which are highly sensitive to the O_2 content in the system (Lastella *et al.*, 2002). Their inactivity depends on an increasing fatty and acetic acids concentration within the environment, which leads to reducing pH value. In a well-balanced system, pH is measured range between 7 and 8(Lastella *et al.*, 2002).

AD usually occurs under the temperature in range of 10-60 °C roughly (Source: http://nongyj.fuyang.gov.cn). There are three AD technologies in terms of different temperature requirement. The production processes in these three AD technologies are basically the same. However, the temperature affects the activity of bacteria participated in the biogas production process, which could influence the retention time and biogas yield. The first one is the digestion occurred under ambient temperature. This AD technology is widely used in rural areas of the developing countries. The digester applied this AD technology does not require a heating system, so it is easy to operate but the biogas output is unstable. For example, in rural China the digester has a lower biogas yield in winter compared to summer.

In northern area the digester usually increase the temperature from a combined greenhouse as mentioned before. Along with the development of biogas technology, mesophilic digestion is widely used in developed countries and some developing countries. Recently, thermophilic digestion has also been develop and used in some joint or large-scale biogas plants due to its advantages.

2.3.2 Feedstock for AD Process

Biogas can be produced from nearly all kinds of biological feedstock types, which are from the primary agricultural sector and different organic waste streams overall society. Feedstock for AD derives from different agricultural, industrial and municipal sources. Agricultural resources include manure (cattle, pig, poultry, etc), energy crops, algal biomass, harvest remains, etc. Industrial resources are from food or beverage processing, dairy, starch industry, sugar industry, biochemical industry, etc.

The largest resource is considered as animal manure and slurries. For instance, more than 1500 million tones of animal manure are produced per year in the EU-27 alone, and more than 65% of these manure are handled as slurry which a liquid mixture of urine, feces, water and bedding material (Holm-Nielsen *et al.*, 2009). Energy crops are another agricultural resource could be used for AD, including grain crops, grass crops and maize, etc, and maize silage is believed to be one of the most promising energy crops for biogas production (Holm-Nielsen *et al.*, 2009). Biomass also can be used for biogas production if containing carbohydrates, proteins, fats, cellulose, and hemicelluloses as main components (Weiland, 2010). Generally, the feedstock type and the digestion system could influence the composition of biogas and biogas yield. Nowadays, in order to obtain a higher biogas yield, most of the agricultural biogas plants digest manure with some additional co-substrates for increasing the content of organic material, particularly in some developed countries. Typical co-substrates are harvest residues such as top and leaves of sugar beets, organic wastes from industries, municipal bio-waste from household, and so on (Weiland, 2010). The biogas yield of every single substrate differs and depends on its origin, content of organic substance and substrate composition.

There are some components, such as inorganic matter like sand, glass, metals, existing in the wasters could cause process failures, like phase separation, sedimentation, flotation etc (Weiland, 2010). Hence the attention must be paid on avoidance of these unwanted components upstream of the digesters. When these components enter the digester, the digestion process will be difficult to control properly. One example is sand. It may exist in the chicken slurry and could cause a reduction of the digester volume because of its rapid sedimentation, then leads to process failure(Steffen *et al.*, 1998). Usually the co-substrates contain some disturbing components. It has to be considered carefully if the wastes contain large amounts of these components, and it could be pre-sorted if possible (Steffen *et al.*, 1998).

The degradation rates of wastes could vary widely because of the different substrate composition. Generally, fats provide the highest biogas yields but require longest retention time because of their poor bioavailability, while carbohydrates and proteins have the faster conversion rates but lower biogas yield (Weiland, 2010). For instance, pig slurry shows a higher biogas yields and methane contents than cow slurry, because pig slurry has a slightly higher fat content (Steffen *et al.*, 1998).

2.3.3　Working Conditions for AD Process

AD is a microbial process that occurs in the absence of oxygen. And in this process, several groups of microbial species degrade the complex organic materials, the producing methane and carbon dioxide ultimately. There are a lot of factors that could affect the amount of biogas produced from a specific digester, such as the substrates (particulate, soluble, biodegradable, etc), the biogas technology (wet or dry fermentation, completely mixed or fixed-bed fermentation), the temperature (mesophilic, thermophilic range), the retention time in the reactor and so on (Gallert and Winter, 2002). Balsam (2006) states that the factors related to working conditions including temperature, loading rate, mixing action, nutrients and management are extremely important to the biogas production.

- **Temperature**

Temperature within the digester is a very important factor affecting the biogas process. In conventional mesophilic digesters, maximum conversion is considered to occur at about 35℃. When temperature decreases 11℃, the biogas production will fall by about 50%. Moreover, keeping the temperature steady is even more important. Variations of as little as 2.8℃ could cause the imbalance of the process by inhibiting methane formation and further cause system failure.

- **Loading rate**

According to the experience, it shows that loading of manure with 6 to 10 percent solids usually works best on a daily basis. The retention time in the digester is in the range of 15 to 30 days.

- **Mixing action**

The mixing action is necessary for the loaded manure to prevent settling and to keep the manure contacting with the bacteria. It can also prevent the scum formation and improve release of the biogas. Mixing the contents of the digester could help to maximize gas production. It can be operated by a mechanical mixer, a compressor, or a closed-circuit manure pump.

- **Nutrients**

The process runs best with C/N ratio between 15:1 and 30:1 (optimally 20:1). And most fresh animal manures meet this requirement and require no additional adjustment. When excessive amounts of exposed feedlot manure become a part of loaded manure, the nutrient imbalance could happen. And crop residues or leaves which both contain high carbon can be added to improve the digester performance.

- **Management**

The digesters need regular and frequent monitoring in order to maintain a steady desired temperature and to prevent the system flow from clogging. If there is no proper management of the digester, a significant decline in gas production could occur and it will require months to correct the problem.

AD process could happen in a wide range of environmental conditions, but the ranges required for optimum condition are narrow. Table.5 shows the optimum condition for AD process. As mentioned, temperature could affect significantly the digestion rate. Although biogas production could also occur at temperatures as low as 10°C, the rate is very slow (Engler *et al.*, 1999). Mesophilic digestion works best under the temperatures of approximate 35°C, while thermophilic digestion works best at approximate55°C.

The values of pH and alkalinity are required in the range of 7-8 and more than 2500mg/L respectively for optimum operation. AD is a quite slow process which typically needs retention time of 15-30 days for mesophilic digestion. And thermophilic digestion is more rapid but more energy is required to heat the digester as mentioned before. Loading rate is based on volatile solids (VS) content of the feed and is usually between 0.15 and 0.35 lb VS/ft3/d for mesophilic digestion.

3.0   BIOGAS APPLICATION

3.1   Multiple Benefits of Biogas Technology

The goal of AD technology is to convert organic wastes into two categories of valuable products which are biogas and the digested substrate, commonly named digestate (Holm-Nielsen *et al.*, 2009). The former is a renewable fuel could be further used to produce green electricity, heat or as vehicle fuel, etc. The latter can be used as an organic fertilizer or be further refined into concentrated fertilizers, fiber products, etc. In this part, I will state and discuss the benefits of AD technology, including the environmental benefits of biogas production, the benefits of digestate used as a fertilizer and the benefits of biogas used as energy source.

3.1.1   Environmental Benefits of Biogas Production

In most of the developing countries, biogas produced from anaerobic digesters is used as fuel substitute for kerosene oil; cattle dung cake, agricultural residues, and firewood (Pathak *et al.*, 2009). Burning of those fuels causes the environmental pollution. Biogas technology is considered to provide the benefits of reducing the emission of GHGs and then mitigating global warming in ways of replacing firewood for cooking, replacing kerosene for lighting and cooking, replacing chemical fertilizers and saving trees from deforestation (Pathak *et al.*, 2009). For example, based on the research performed by Pathak *et al.* (2009) in India, a family size biogas plant substitutes 316 L of kerosene, 5,535 kg firewood and 4,400 kg cattle dung cake as fuels every year. It means a family size biogas plant reduces NOx of 16.4 kg, SO2 of 11.3 kg, CO of 987.0 kg and volatile organic compounds of 69.7 kg per year.

Methane is a major GHGs in the world, with a global warming potential (GWP) of 25 times higher than CO_2. Methane emissions could happen in any anaerobic processes with organic materials. Current disposal practices for manure slurry and food residues lead to methane released through natural processes (Klingler, 2000). It has been estimated that emission from agriculture accounts for 33% of the global greenhouse effect (Klingler, 2000). About 7% is from animal excrement which roughly equals to 20-30 million tones of methane every year (Klingler, 2000). Through AD technology for treatment of animal excrement these gases can be used as a fuel and a well-managed AD scheme could maximize methane generation, but not release any gas to the atmosphere. Moreover, AD technology provides the environmental benefits by using renewable energy instead of fossil fuel to reduce CO2 emissions and mitigate other environmental degradations. For instance, in developing countries the small agricultural biogas plants contribute to reduce the use of forest resources for household energy purposes, thereby slowing down deforestation, soil degradation and easing the problems like flooding or desertification.

Nitrous oxide emissions are significantly harmful to the climate change due to its high GWP of 320. Recent research states that AD of animal waste largely reduces nitrous oxide emissions because it helps to avoid emissions from storage of animal waste, reduce application of inorganic nitrogen fertilizer and avoid emissions from production of nitrogen fertilizer, etc (Klingler, 2000). Besides the effects mentioned above, there are numbers of additional environmental benefits provided by AD technology (Source: http://www.adnett.org/).

- **Energy balance**

A well designed and operated AD plant can achieve a better energy balance if taking emissions from transport operations into account than many other forms of energy production. The energy balance depends on the amount of energy consumed for producing energy.

- **Wastewater treatment**

In some countries, in particular southern European countries, biogas technology has been considered as a wastewater treatment system because their manure contains very low dry matter contents and is treated similarly to wastewater. It has several environmental impacts.

Firstly, anaerobic system needs much less land compared with aerobic systems for wastewater treatment. So AD could contribute to preserving valuable land resources. Secondly, it has positive energy balance because anaerobic system needs little process energy compared to generated energy.

- **Recycling nutrients**

The products for AD plants, including liquid fertilizer and fibre, can reduce the demands for synthetic fertilizers within an overall fertilizer program if properly applied.

- **Reducing land and water pollution**

Inappropriate disposal of animal slurries could result in land and ground water pollution. AD technology creates an integrated management system which reduces the possibility of this problem happening.

- **Supporting Organic Farming**

AD has the potential to support Organic Farming when used as part of a closed loop. Generally organic fertilizer contains weed seeds and microorganisms resulting in pests. They cause the use of herbicides and pesticides in farming system. However, AD process could reduce the ability of seeds to germinate and minimizes the survival of microorganisms. So the use of digestate from AD as a fertilizer could contribute to organic farming due to this effect.

- **Reducing odour**

For many farmers solving the problem of odour is an important reason to install a biogas plant. AD for manure treatment allows farmers to remove manure which causes the odour complains. It is reported that AD could reduce the odour from farm slurries and food residues by up to 80%.

3.1.2 The Benefits of Digestate Used as Fertilizer

Along with the biogas produced, AD also transforms the added feedstock into digestate that can be used as a fertilizer which is high in nitrogen, potassium and phosphorus contents. The digestate can be stored then used in farmlands for crop production at an appropriate time without further treatment. Besides, it can be separated to produce fibre and liquor. The fibre can be sold or used as a good fertilizer or a soil conditioner, while the liquor contains various nutrients and could be used as a liquid fertilizer which could be sold or used on-site. Fig.7 shows mass balance for AD process. Usually 7-25% of Fibre and 75-95% of Liquor are produced

The digestate almost remains all the non-degradable substances from the original feedstock as well as all plant nutrients. The nutrient content of digested slurry depends on which type of feedstock (manure, co-substrates, etc) is digested. Moreover, AD process of manure or other organic biomass could transform part of organic bound nutrients to a mineral form (Ørtenblad, 2000). This effect is very important for nitrogen. In AD process, part of the organic nitrogen such as proteins is released as ammonium (Ørtenblad, 2000).

Ammonium is readily available for the crops when it is applied to the fields (Ørtenblad, 2000). It also helps to reduce the need for using additional mineral nitrogen fertilizers. So the digestate from anaerobic fermentation is considered as an improved and valuable fertilizer which could substitute mineral fertilizer due to the increased availability of nitrogen to crops. In addition, anaerobic treatment minimizes the survival of pathogens from the feedstock, which is important for the digestate used as a fertilizer (Ørtenblad, 2000).

The reuse of the digestate from AD as fertilizer presents a sustainable way to control and direct nutrients in society. If it is implemented widely, it will be possible to recover the broken nutrient cycle between the productive soils of the countryside and the consuming people of the cities nowadays, which could facilitate the reduction of the use of mineral fertilizers (Lantz *et al.*, 2007).

3.2   The Utilization and Application of Biogas

Biogas is an ideal energy source and suitable for practically all the various fuel requirements in the household, agriculture and industrial sectors. However, the different standards of gas quality are required by the individual gas utilization, which make purification and upgrading of the gas necessary. There are various biogas utilization purposes, including: production of heat or steam (the lowest value chain utilization); industrial energy source for heat, steam, electricity, cooling, etc; electricity production with CHP; upgraded and used as vehicle fuel; upgrading and injection into the natural gas grids; fuel for fuel cells, etc.

In the developing countries the most common utilization of biogas from small-scale plants is on-farm application, including cooking, lighting, heating (space heating, water heating, and grain drying), cooling, etc. In most cases, the equipment designed for burning natural gas requires slightly modifications to fit the different burn characteristics of biogas (Balsam, 2006).

In a number of industrial applications, biogas can be used in small-scale industrial operations for direct heating applications such as in scalding tanks, drying rooms and in the running of internal combustion engines for shaft power needs (Akinbami, 2001). It could also be used for steam production.

Biogas produced by co-substrates of manure with energy crops or harvesting residues may contain H2S whose level is in the range of 100-3,000ppm (Weiland, 2010). However, the CHP station for biogas utilization requires level of H2S below 250ppm, in order to avoid excessive corrosion and expensive deterioration of lubrication oil (Weiland, 2010). Today biological desulfurization is a main method for removal of H2S. Small-scale biogas plants are widely used for CHP in decentralized on-farm units. Typical output from CHP station based on biogas is about 2/3 thermal and 1/3 electricity at 80-90% efficiency (Poeschl *et al.*, 2010).The generated heat is commonly used for heating of digesters and the local residential houses and animal stalls, but it also could be used for heat transmission to public buildings, grain drying, production of animal feed, and drying of wood fuel. Nevertheless, heat transmission causes heat losses in the range of 3.5-20% dependent on the transmission distance (Poeschl *et al.*, 2010). Electricity generated from CHP could be sold to independent energy supplies.

In some EU countries, biogas is scrubbed of carbon dioxide and other impurities to generate a CH4-enriched biogas which is 95–98% CH4 (Murphya *et al.*, 2004). This CH4-enriched biogas could be used as vehicle fuel. For example, Volvo has developed a bi-fuel car, which runs on petrol and biogas (Murphya *et al.*, 2004).This offers flexibility to the consumer and could maximize utilization of the biogas. A remarkable example of biogas utilization as vehicle fuel is Sweden. It is reported that the market for such biogas utilization has been increasing rapidly in the last decade, and today there are 15,000 vehicles based on upgraded biogas in Sweden (Persson *et al.*, 2007). It is predicted there will be 70,000 vehicles running on biogas supplied from 500 stations, by the year of 2010-2012 (PERSSON *et al.*, 2007).

In addition, CH4-enriched biogas could also be introduced into the natural gas-grid to support a series of biogas service stations. In some EU countries like Germany, Sweden and Switzerland have developed the quality standards for biogas injection into the natural gas-grid (Weiland, 2010). Upgrading and injecting biogas into the natural gas-grid is an efficient way of integrating the biogas into the energy sector. Since biogas cannot always be used nearby the production plants, injecting upgraded biogas into the natural gas-grid offers the opportunities to transport and use biogas in the larger energy consumption areas, where the population is intensive (Holm-Nielsen *et al.*, 2009).

These two applications of biogas which are utilization as vehicle fuel and injection into the gas-grid have become more and more important because the gas can be used in a

relatively energy efficient way.

4.0   CONCLUSION

Biogas is widely in use all over the world, but the status varies among the different continents. In Africa most of biogas plants are small-scale but the large-scale AD technology is still under developed. The dissemination of biogas plants is still difficult in Africa. Some Asian countries such as China and India have built a large number of biogas plants. Millions of people, in particular farmers, have benefited from the biogas technology. More sophisticated plants are found in developed countries.

Biogas technology influences the energy consumption and utilization by replacing various fuels and saving energy consumption. It produces renewable energy by using local input. It significantly benefits the environment in term of reduction of GHGs emissions, and it benefits the agricultural practice. Biogas technology represents a sustainable way to produce energy for rural household, particularly in developing countries. Other factors need to be considered, such as biogas digester type (if requiring additional energy input), feedstock type, water resource supply and storage tank for digestate (if household has not farmland) when doing analysis and calculations in other rural area. China has made a big achievement in developing and implementing biogas technology in both rural household and large-scale biogas plants. The reasons include (a).plenty of feedstock. Along with the development of livestock industry, more manure could be collected as feedstock for biogas production. In addition, the plentiful crop yield also ensures the supply of agricultural residues for biogas production.

(b).strong state financial subsidy.

These reasons may provide some experience to other developing countries. Certainly, the developed countries have more advanced biogas technologies in terms of biogas production and utilization, and these technologies can also become relevant in developing countries.

REFERENCES

Akinbami, M. O. I. B., T.O. Oyebisi B, I.O. Akinwumi B, O. and Adeoti, C. (2001). Biogas

energy use in Nigeria current status, future prospects and policy implications.pdf.

Renewable and Sustainable Energy Reviews, 5.

Amigun, B. and Blottnitz, H. (2007). Investigation of scale economies for African biogas

installations. Energy Conversion and Management, 48, 3090-3094.

Amigun, B. and Blottnitz, H. (2010). Capacity-cost and location-cost analyses for biogas

plants

in Africa. Resources, Conservation and Recycling, 55, 63-73.

Balsam, J. (2006). Anaerobic Digestion of Animal Wastes Factors to Consider.pdf. ATTRA –

National Sustainable Agriculture Information Service.

Birkmose, T. (2007). Digested manure is a valuable fertilizer. University of Southern Denmark,

Esbjerg, Denmark.

Engler, C. R., Jordan, E. R., Mcfarland, M. J. and Lacewell, R. D. (1999). Economics and

Environmental Impact of Biogas.pdf. In: Proceedings of the 1999 Texas Animal Manure

Management Conference., Texas A&M University. CollegeStation, TX, USA.

Renewable Energy - Targets by 2020, European Commission. 2011:

Gallert, C. and Winter, J. (2002). Solid and liquid residues as raw materials for biotechnology.

Die Naturwissenschaften, 89, 483-96.

Gautam, R., Baral, S. and Herat, S. (2009). Biogas as a sustainable energy source in Nepal:

Present status and future challenges. Renewable and Sustainable Energy Reviews, 13, 248-252.

Holm-Nielsen, J. B., Al Seadi, T. and Oleskowicz-Popiel, P. (2009). The future of anaerobic digestion and biogas utilization. Bioresource technology, 100, 5478-84.

Klingler, B. (2000). Environmental Aspects of Biogas Technology. in: AD-Nett (ed): AD: Making energy and solving modern waste problems, Ortenblad H Herning municipal utilities, Denmark, pp.53-65.

Lantz, M., Svensson, M., BJ Rnsson, L. and B Rjesson, P. (2007). The prospects for an expansion of biogas systems in Sweden—Incentives, barriers and potentials. Energy Policy, 35, 1830-1843.

Lastella, G., Testa, C., Cornacchia, G., Notornicola, M., Voltasio, F. and Sharma, V. K. (2002). Anaerobic digestion of semi-solid organic waste biogas production and its purification.pdf. Energy Conversion and Management, 43.

Murphya, J. D., Mckeoghb, E. and Kielyb, G. (2004). Technical economic environmental analysis of biogas utilization.pdf. Applied Energy, 77.

Omer, A. M. and Fadalla, Y. (2003). Biogas energy technology in Sudan.pdf. Renewable Energy, 28.

Ortenblad, H. (2000). The use of digested slurry within agriculture. Herning Municipal Utilities, Denmark, http://www.adnett.org/.

Parawira, W. (2009). Biogas technology in sub-Saharan Africa: status, prospects and constraints.

Reviews in Environmental Science and Biotechnology, 8, 187-200.

Pathak, H., Jain, N., Bhatia, A., Mohanty, S. and Gupta, N. (2009). Global warming mitigation potential of biogas plants in India. Environmental monitoring and assessment, 157, 407-

18.

Persson, M., Jönsson, O. and Wellinger, A. (2007). Biogas upgrading to vehicle fuel standards and grid injection. IEA Bioenergy, Task 37-Energy from Biogas and Landfill Gas.

Poeschl, M., Ward, S. and Owende, P. (2010). Prospects for expanded utilization of biogas in Germany. Renewable & Sustainable Energy Reviews, 14, 1782-1797.

Steffen, R., Szolar, O. and Braun, R. (1998). Feedstocks for Anaerobic Digestion. in: AD-Nett. http://www.adnett.org/.

Van Nes, W. J. and Nhete, T. D. (2007). Biogas for a better life. Renewable Energy World, July–August 2007.

Weiland, P. (2010). Biogas production: current state and perspectives. Applied microbiology and

biotechnology, 85, 849-60.

YOUR KNOWLEDGE HAS VALUE

- We will publish your bachelor's and master's thesis, essays and papers

- Your own eBook and book - sold worldwide in all relevant shops

- Earn money with each sale

Upload your text at www.GRIN.com and publish for free